Problem Solving Strategies (4-6)

Portsmouth, NH
800-596-3175
www.onboardacademics.com
ISBN: 978-1-63096-076-6

OnBoard Academic's books are specifically designed to be used as printed workbooks or as on-screen instruction. Each page offers focused exercises and students quickly master topics with enough proficiency to move on to the next level.

OnBoard Academic's lessons are used in over 25,000 classrooms to rave reviews. Our lessons are aligned to the most recent governmental standards and are updated from time to time as standards change. Correlation documents are located on our website. Our lessons are created, edited and evaluated by educators to ensure top quality and real life success.

Interactive lessons for digital whiteboards, mobile devices, and PCs are available at www.onboardacademics.com. These interactive lessons make great additions to our books.

You can always reach us at customerservice@onboardacademics.com.

Problem Solving Strategies
Inverse Operations

Key Vocabulary

inverse operations

work backwards

Chocolate Smoothie Problem

Mia made a chocolate smoothie with 12 oz of milk and some chocolate ice cream. She filled four 6-oz glasses and there were 3 oz left over.

How many ounces of chocolate ice cream did she use?

12 + ? = 27

27 – 12 = 15

Mia used 15 oz of ice cream.

Work backwards using inverse operations to solve problems.

Addition and Subtraction are Inverse Operations

3	+	5	=	8
	−		=	

Multiplication and division are inverse operations.

$4 \times 6 = 24$

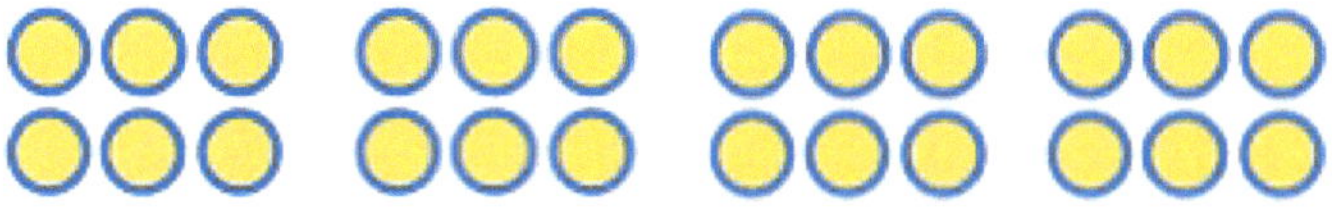

$24 \div 4 = 6$

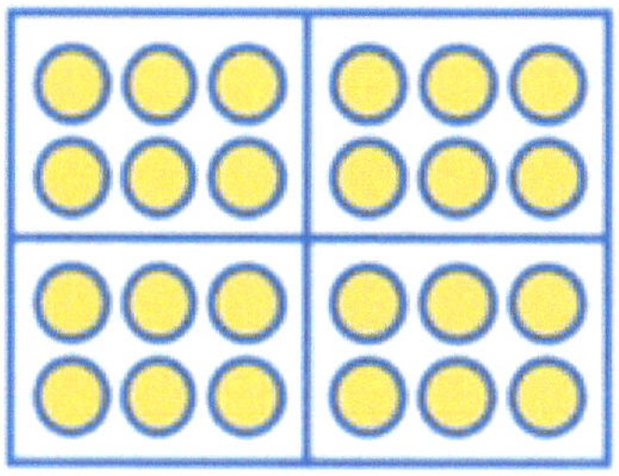

$\boxed{4} \times \boxed{6} = \boxed{24}$

$\boxed{} \div \boxed{} = \boxed{}$

Use inverse operations to complete these operations.

15	−	9	=	6	INVERSE		+		=	
15	−	9	=	6		6	+	9	=	15
	+	9	=		INVERSE	23	−		=	
	x	8	=	56	INVERSE		÷		=	7
	÷	4	=		INVERSE	12	x		=	

How old is James?

Complete the flowchart below by adding x3 and ÷3 to calculating James age.

James' Dad is 54 years old. If James starts with his own age, adds 9, then multiplies by 3, he will get to his Dad's age. How old is James? Solve using a flowchart.

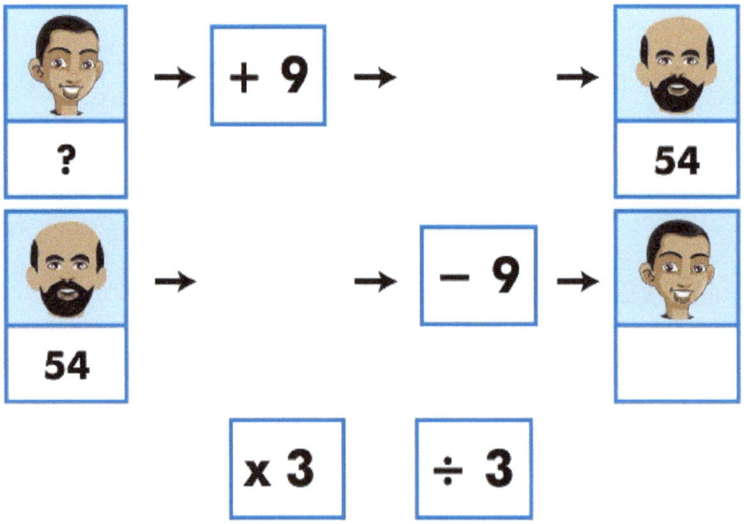

Work backwards to solve.

Brian and Jenna are sharing a bowl of pistachio nuts while watching a game of football. In the first half, Brian takes 7 nuts and Jenna takes 3 nuts. At the start of the second half, Brian takes half of the remaining nuts, while Jenna takes 10 nuts. If there are 15 nuts left for the 4th quarter, how many nuts did they have to begin with?

?	→		→		→		→		→	= 15

	→		→		→		→		→	

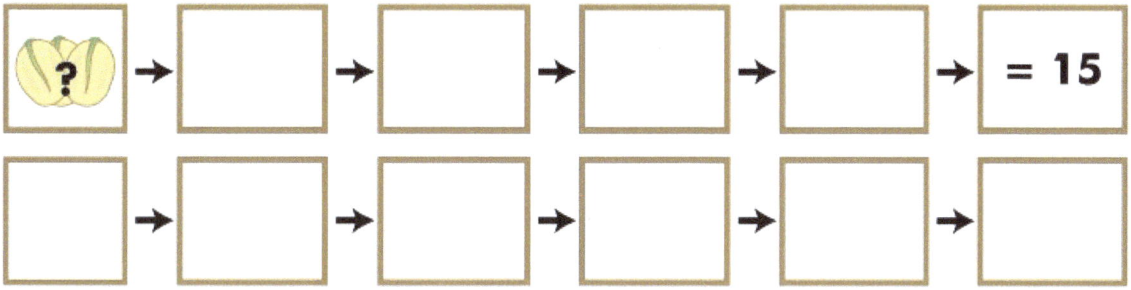

Name______________________________

Inverse Operations Quiz

1. True or false? Addition and multiplication are inverse operations.

2. Which equation uses inverse operations to $42 - 11 = 31$?

   A. $42 \times 11 = 31$

   B. $31 + 11 = 42$

   C. $42 \div 11 = 31$

   D. $31 - 11 = 42$

3. Which digit is missing in these inverse equations:
   $? \times 6 = 72$ $72 \div 6 = ?$

4. Ingrid's hamster has pups. A friend takes $\frac{1}{2}$ of the pups, but returns three, leaving Ingrid to care for 15. pups. How many pups were in the litter?

Problem Solving Strategies
Patterns and Tables

Key Vocabulary

pattern

table

rule

sequence

Extend a Pattern

Circle the correct symbol to extend the pattern.

Find and extend these numerical patterns.

Alison is hand painting a border around her bedroom wall. Extend the pattern.

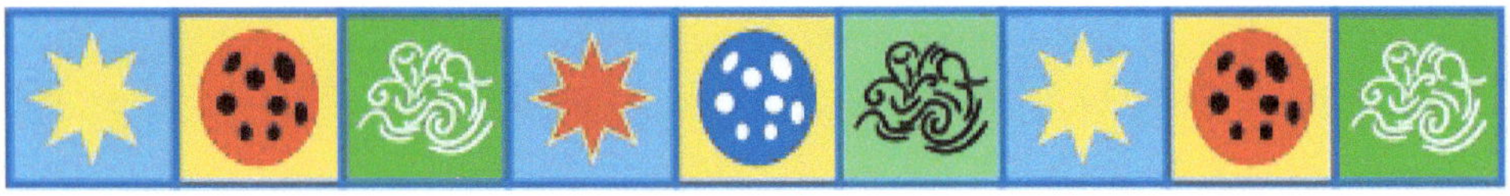

1 60, 80, 100, 120, ____ , ____

2 2, 5, 4, 7, 6, ____ , ____

3 5, 10, ____ , ____

Finding and extending a pattern.
Complete the patterns below.

3	6	9	12			Add 3 RULE
+ 3	+ 3	+ 3				

100	90	80	70			RULE

4, 8, 7, 11, 10, 14, 13				RULE
	+ 4	− 1	+ 4	

Organizing a Data Table

Read the information provided about James reading progress. Creating a data table is a good strategy to solve this problem. Discover the data table that has been created and discover how it is used to solve the problem.

James is reading a 400-page novel, but is finding it a little hard going. He is only reading 80 pages a month!
How many months will it take him to finish the novel?

Number of pages read by James each month

Month #	1	2	3	4	5	6	7	8
Total # Pages	80							

title

heading

Number of pages read by James each month

Month #	1	2	3	4	5
Total # Pages	80	160	240	320	400

heading

Tables can help you to organize data

Create a table and solve.
All the elements are provided below.

Fernando scored 80 on his first math quiz. On his second math quiz, he scored 84. He was excited to receive a score of 88 on his third quiz. If he continues to improve at this rate, on which quiz will he score 100?

88 90 9 84 100 Score

92 7 80 3 10 5 4 94 2

8 6

96 1 Quiz 98

Name______________________________

Patterns and Tables Quiz

1 True or false, the rule for this pattern is add 4?
4, 8, 16, 32...

2 What is the rule for this pattern? 5, 8, 6, 9, 7, 10, 8...

A. Add 3

B. Subtract 2

C. Add 3, subtract 2

D. Subtract 3, add 2

# Boxes	S & H
1	$6
2	$11
3	$15
4	$18

Table 1

3 What is the shipping and handling (S & H) cost, in dollars, for shipping 5 boxes? (Table 1 above)

4 If Owen earns $18 every week, and saves $8, how many whole weeks will it take him to save enough to buy a $60 digital player?

Addition & Subtraction

Key Vocabulary

sum

difference

regrouping

ones

tens

hundreds

Regroup using the fewest number of 10 base blocks.
Draw your answer in the boxes provided.

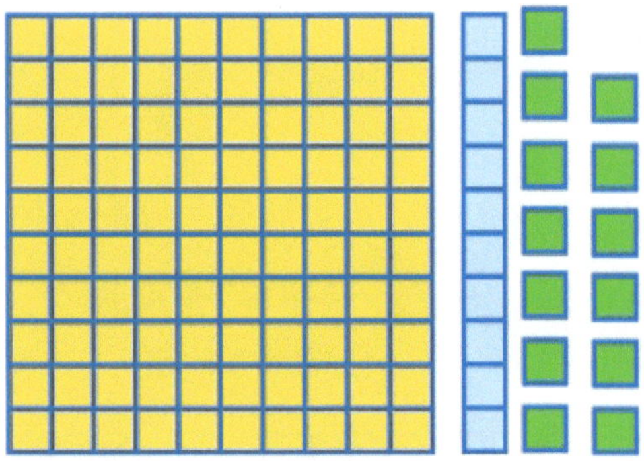

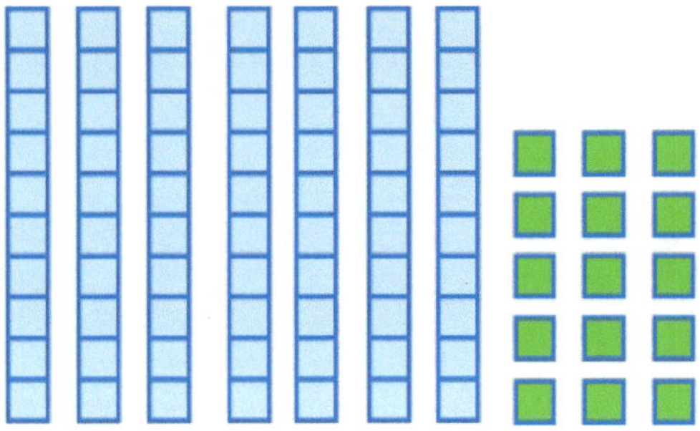

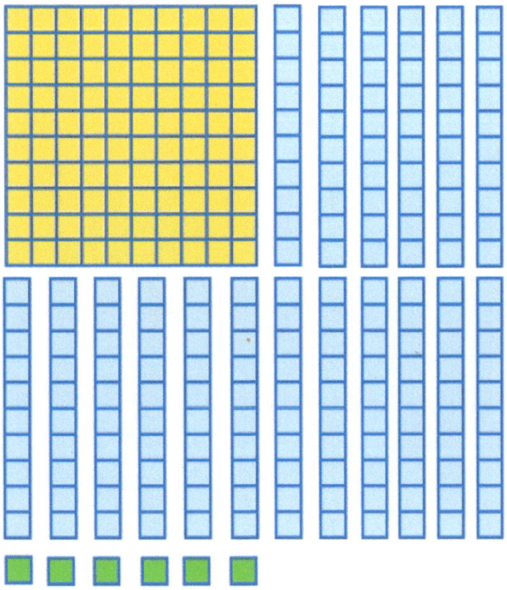

Model the solution to this addition problem.

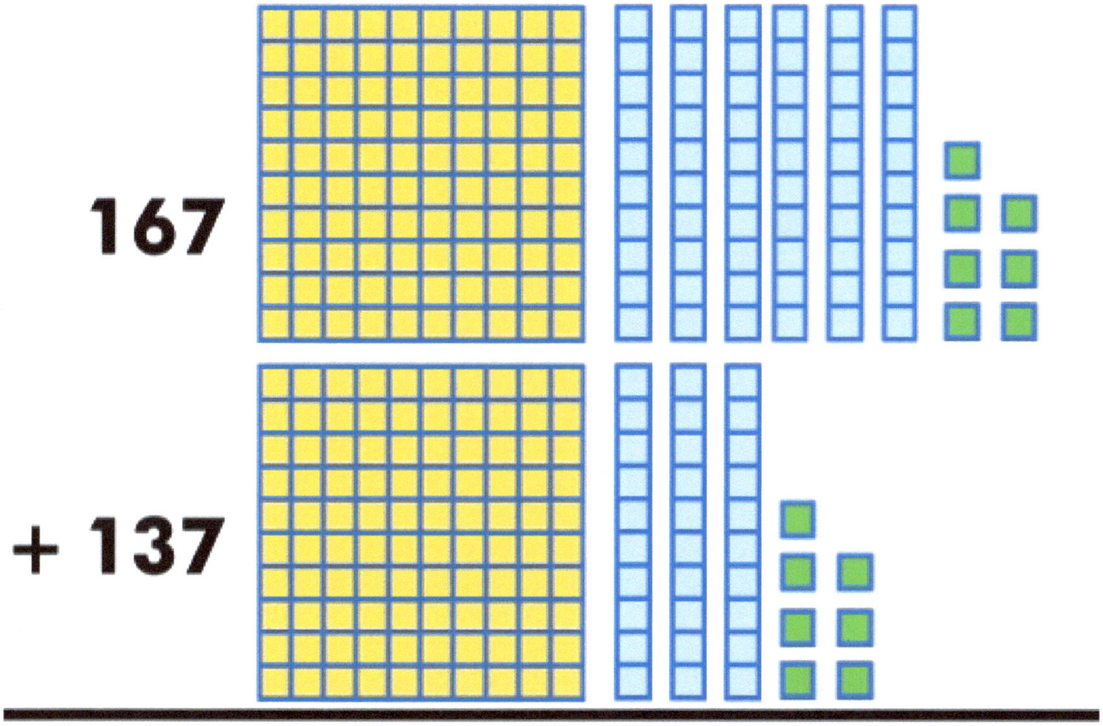

Adding with Regrouping

$$\begin{array}{r} 434 \\ +\ 387 \\ \hline \end{array}$$

1

Line up the place values

$$\begin{array}{r} 434 \\ +\ 387 \\ \hline \\ \hline \end{array}$$

2

Add the ones:
4 + 7 = 11
Regroup: trade 10 ones for 1 ten

3

Add the tens:
3 + 8 +1 = 12
Regroup: trade 10 tens for 1 hundred

4

Add the hundreds:
4 + 3 + 1 = 8
Solution: 821

Model the solution to this subtraction problem.

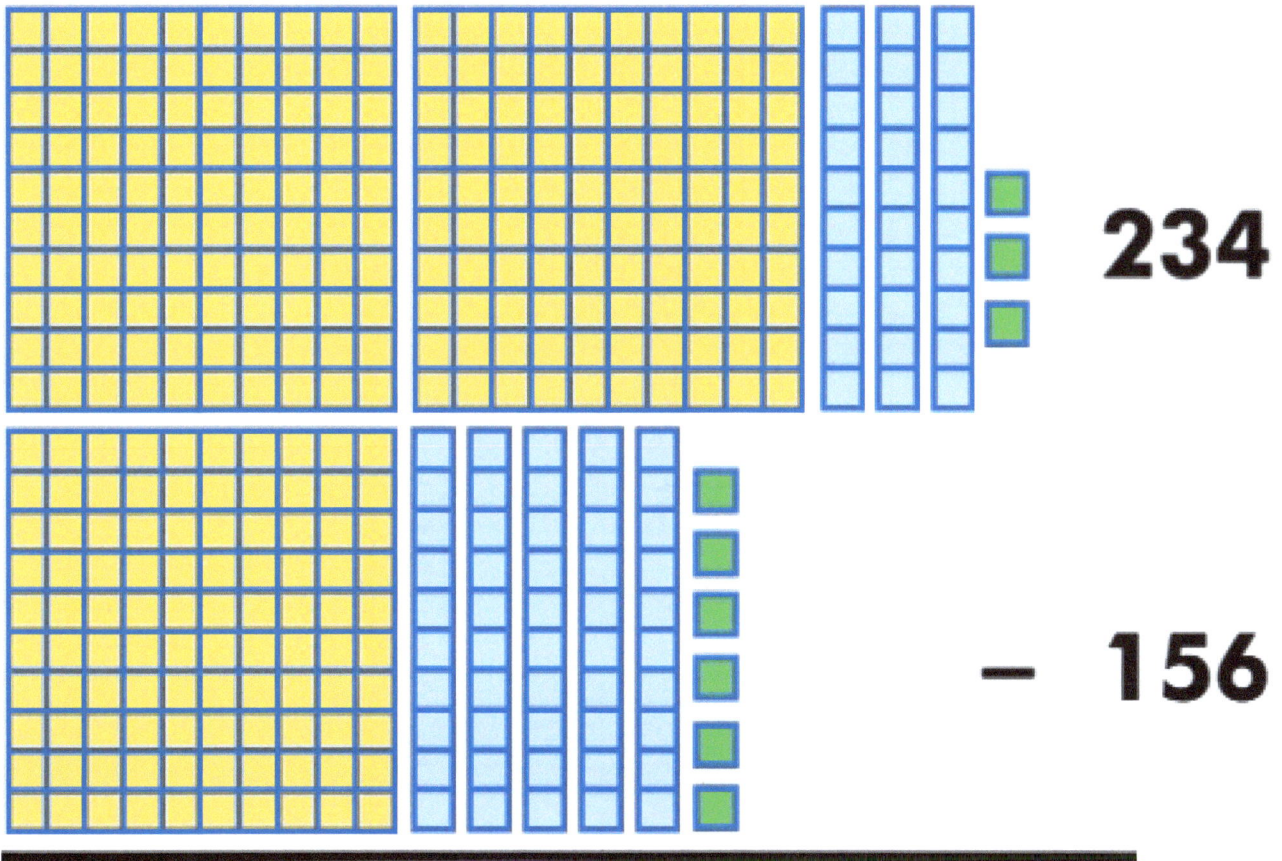

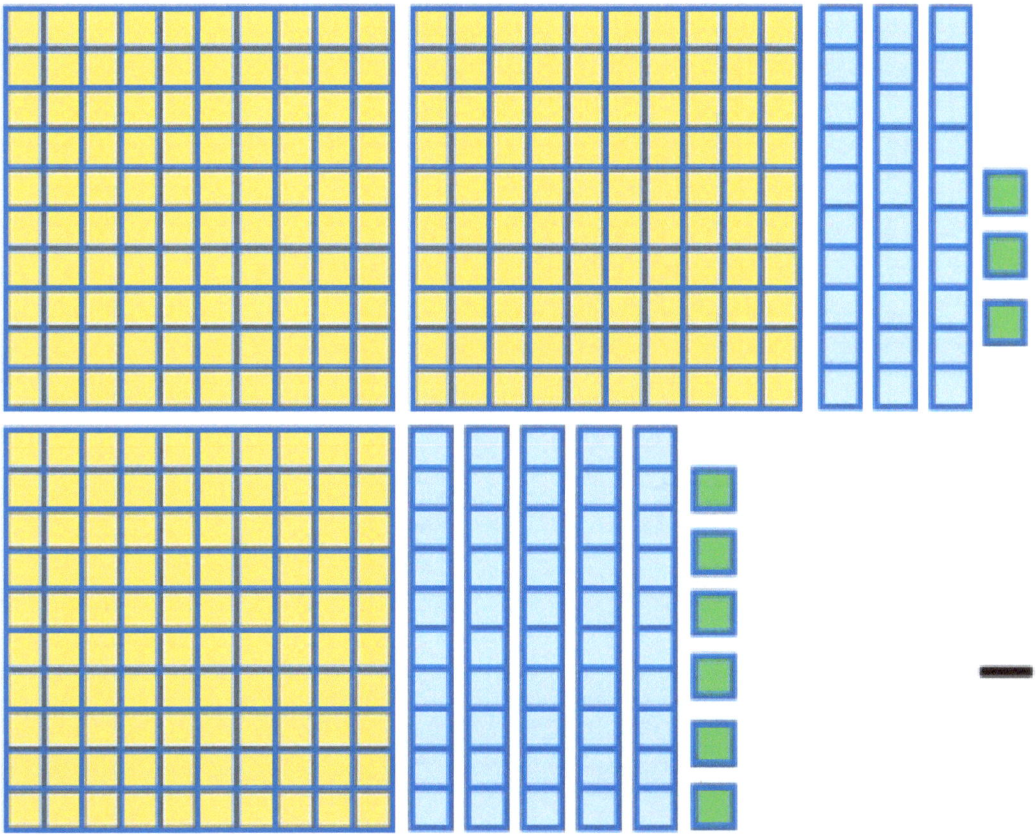

$$\begin{array}{r} 234 \\ -\ 156 \\ \hline \end{array}$$

Subtracting with Regrouping

$$\begin{array}{r} 7\ 1\ 3 \\ -\ 3\ 4\ 5 \\ \hline \end{array}$$

1

Line up the place values

$$\begin{array}{r} 7\ 1\ 3 \\ -\ 3\ 4\ 5 \\ \hline \end{array}$$

2

Subtract the ones:
regroup from the tens
13 – 5 = 8

3

Subtract the tens:
regroup from the hundreds
10 – 5 = 5

4

Subtract the hundreds:
6 – 3 = 3
Solution: 368

Practice addition and subtraction.

1.
$$\begin{array}{r} 373 \\ +\ 248 \\ \hline \end{array}$$

______

2.
$$\begin{array}{r} 621 \\ -\ 234 \\ \hline \end{array}$$

______

3.
$$\begin{array}{r} 503 \\ -\ 117 \\ \hline \end{array}$$

______

4.
$$\begin{array}{r} 625 \\ +\ 789 \\ \hline \end{array}$$

______

Can you solve these word problems?

During a school-wide reading contest, the fourth graders read 556 books. The fifth graders read 478 books.

How many books did the two grades read altogether?

How many more books did the fourth graders read compared to the fifth graders?

Which operation should you use for each problem?

Name______________________________

Addition and Subtraction Quiz

1. True or false? To find the sum of two numbers, you subtract them.

2. 312 – 197 = ?

   A. 117

   B. 285

   C. 115

   D. 125

3. 478 + 593 = ?

4. Michael has collected 397 baseball cards. Kevin has collected 512 baseball cards. How many more baseball cards has Kevin collected?

Judging Reasonable Answers

Key Vocabulary

estimate

nice number

reasonable answer

Use estimation to find "reasonable" answers.

Place check mark in the box if it is a reasonable answer and an X if it is not a reasonable answer.

"I scored 78, 80, and 82 in math. My average was 95!" ☐

"I gave 8 sticks of gum to all 17 people in my class. That's about 150 sticks!" ☐

"On my 8th birthday, I added up my age in days. I was 740 days old!" ☐

"I subtracted 62 from 100 and came up with an answer of 38." ☐

Make and estimate using nice numbers.

Round the numbers in red by drawing them the number line to the nearest 10.
This will give you “nice numbers” that are easy to add.

"Nice numbers" are easy to add and subtract.

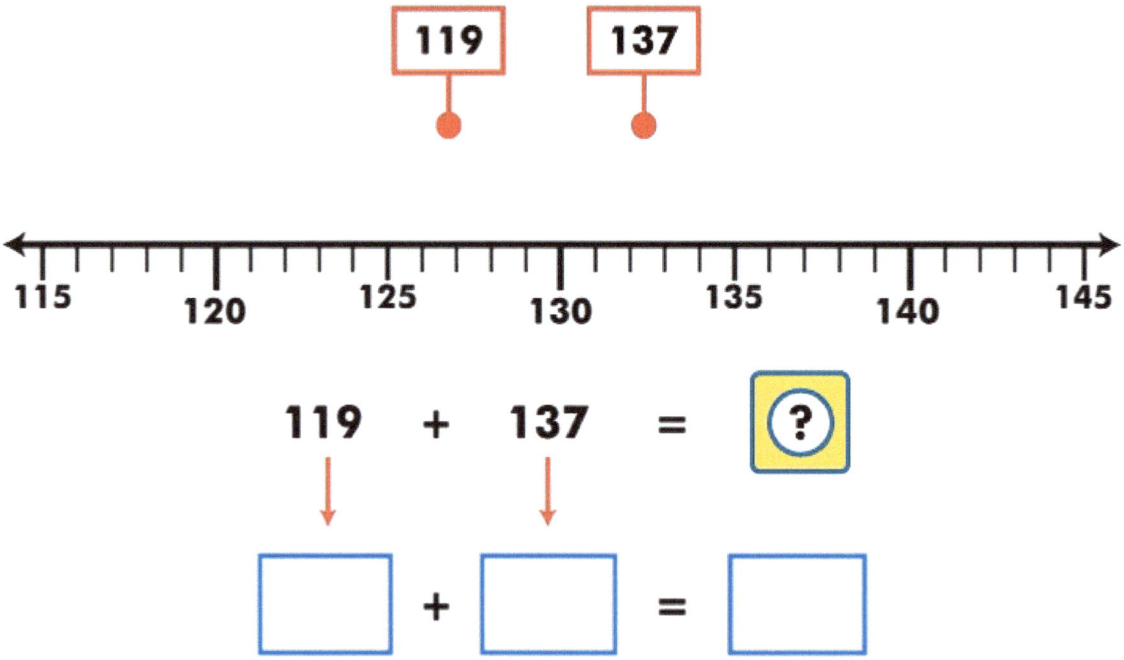

Use number friends to calculate the total points scored by Owen.
Number friends are numbers that when added together make nice numbers. The number friends are color coordinated below.

Points scored by Owen

Game	Points
1	15
2	10
3	40
4	25
5	35
6	25
TOTAL	

Use estimation to judge reasonable answers.
Fill in the blanks with R or N.

5 x 62 = 30 ☐	765 – 567 = 450 ☐
43 + 28 + 97 = 130 ☐	3 x 78 = 234 ☐
49 – 24 + 49 = 25 ☐	812 ÷ 9 = 90 ☐

R easonable   **N** ot reasonable

Owen, Fernando and Mia each want to spend about $6 on lunch.
Can you select lunch items for each person so that your estimate is about $6

Make reasonable estimates below.

There are 345 girls on 5 buses.	girls per bus
Grandma drove for $5\frac{1}{2}$ hours at 37 mph.	total no. miles
Jack paid $3.50 for 5 bottles of water.	cost per bottle
It's 3,579 miles to Miami. We've driven 2,856 miles.	no. miles to go

Name____________________________

Judging Reasonable Answers Quiz

1. **True or false? If you drive for 10 hours at 53 mph, you will have driven more than 500 miles.**

2. **A golfer shoots 68, 72, and 70. What is his average?**

   (A) **68**

   (B) **69**

   (C) **70**

   (D) **71**

3. **What's the best estimate for 62 x 29 = ?**
   1,800   1,200   1,500   2,000

4. **What's the best estimate for 789 ÷ 8 = ?**
   50   80   100   150

Greatest Common Factor

Key Vocabulary

factor

greatest common factor (GCF)

Find the factors of 24 and 32.

If the factor is just for 24 write it in the yellow space, if its just for 32 write it in the blue space and if the factor is for both 24 and 32 write it in the green space.

1

2

3

4

5

6

7

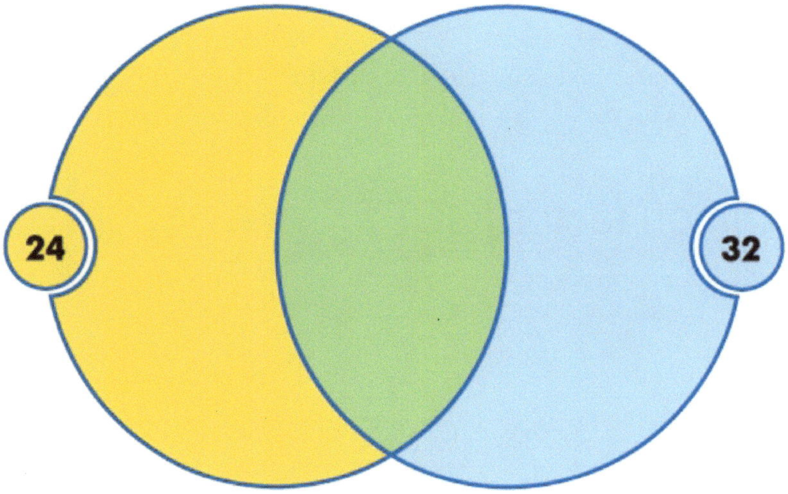

8 9 11 12 14 16 18 24 26 32

The numbers in the green space are the common factors..

What is the greatest common factor (the largest)? Circle it.

Finding the GCP of 30 and 10.

1 Find the factors of 30	1	2	3	5	6	10	15	30
2 Find the factors of 10	1	2	5	10				
3 Find the common factors								
4 Find the greatest common factor (GCF)								

Finding the GCF using prime factors.

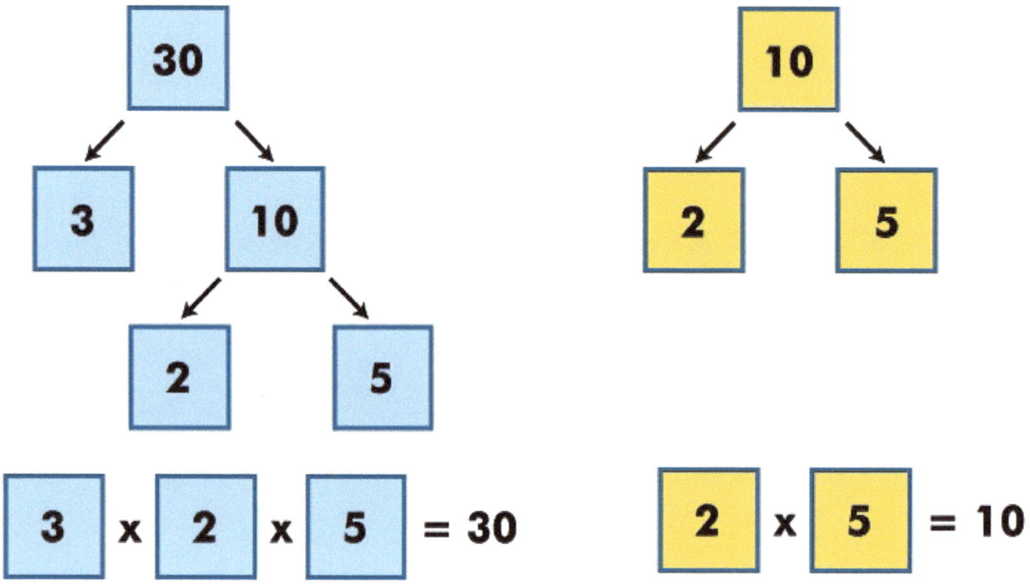

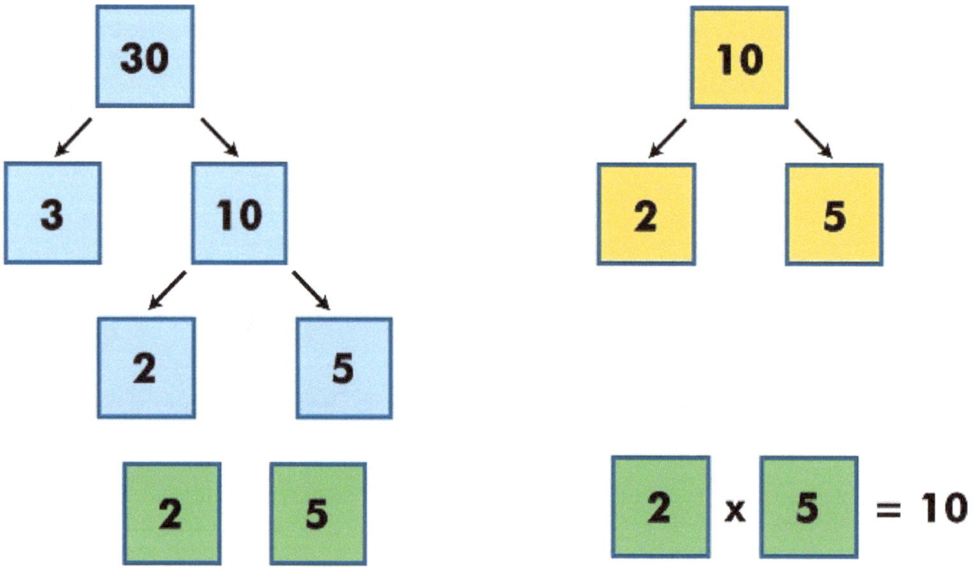

The GCF is the product of the common prime factors

Use the prime factors to find the GCF of 18 and 24.

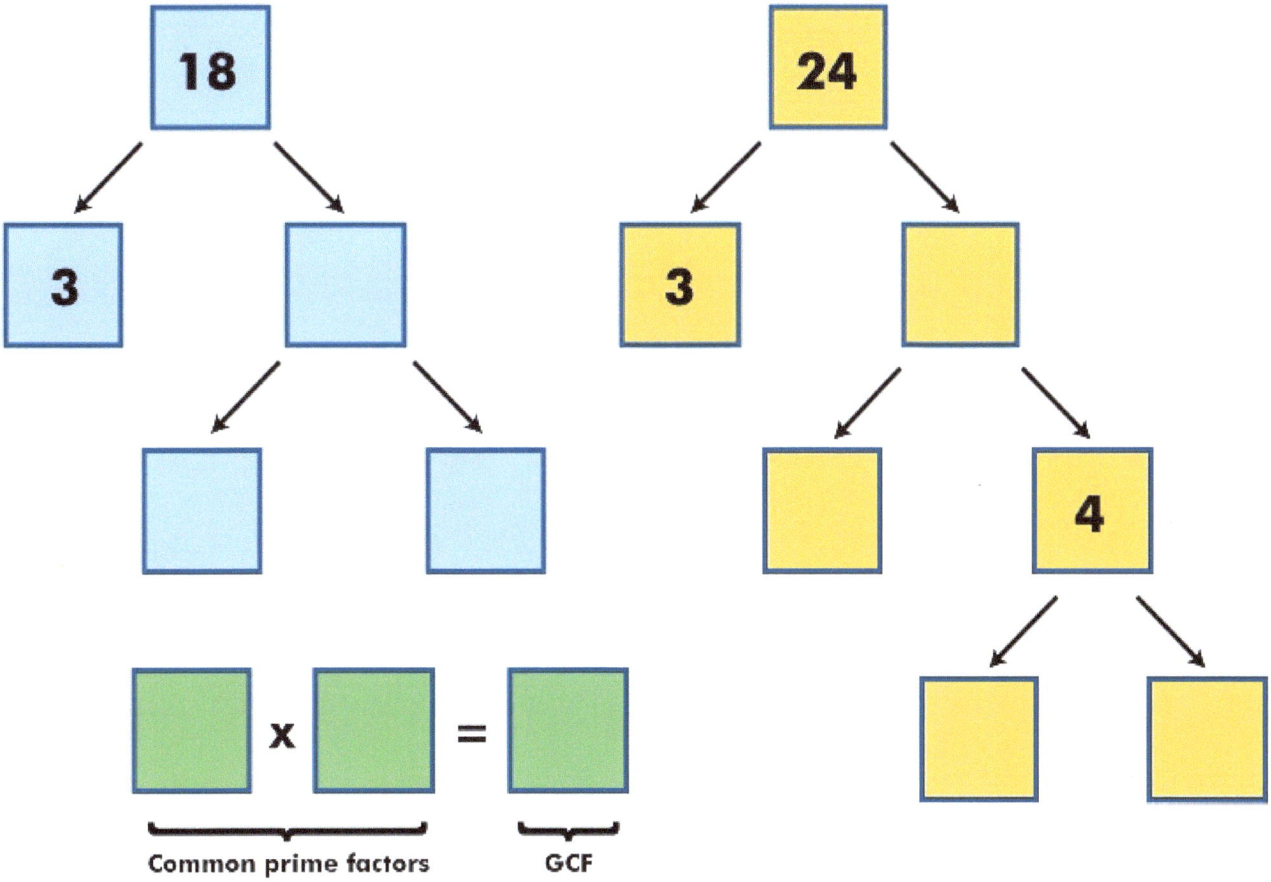

Solve this word problem using prime factors.

Sanjeev has a part-time job stacking shelves. His boss asks him to stack 42 boxes of *Iron-Bran* and 56 boxes of *Fruity Fiber*. He must stack them in such a way that each row has the same number of boxes, but he can not mix boxes in the same row.

What is the greatest number of boxes that can be in each row?

Name________________________________

Greatest Common Factor Quiz

1. The GCF of 8 and 16 is 4. True or False?

2. The common prime factors of 8 and 16 are:
   - A. 2 x 2 x 2
   - B. 2 x 2
   - C. 2 x 4
   - D. 2 x 2 x 2 x 2

3. What is the GCF of 9 and 42?

4. What is the GCF of 12, 16 and 24?

Multiples & Least Common Multiples (LCM)

Key Vocabulary

factor

multiple

least common multiple (LCM)

Multiple or Factor?

The multiple of a number is the result of multiplying that number by another whole number, e.g. 10 is a multiple of 5.

Write the correct description in the box provided.

16 is a	☐	**of 32**
15 is a	☐	**of 5**
8 is a	☐	**of 2**
2 is a	☐	**of 4**

Multiple

or

Factor

Multiples and Common Multiples

The first five multiples of 8:

The first six multiples of 4:

The first three multiples of 12:

Which number is common to all three lists?

We call this number a common Multiple. This number is also the least common multiple of 8, 4 and 12. This is sometime referred to by its initials LCM.

Find the multiples of 6 and 4 by drawing a circle around the multiple.

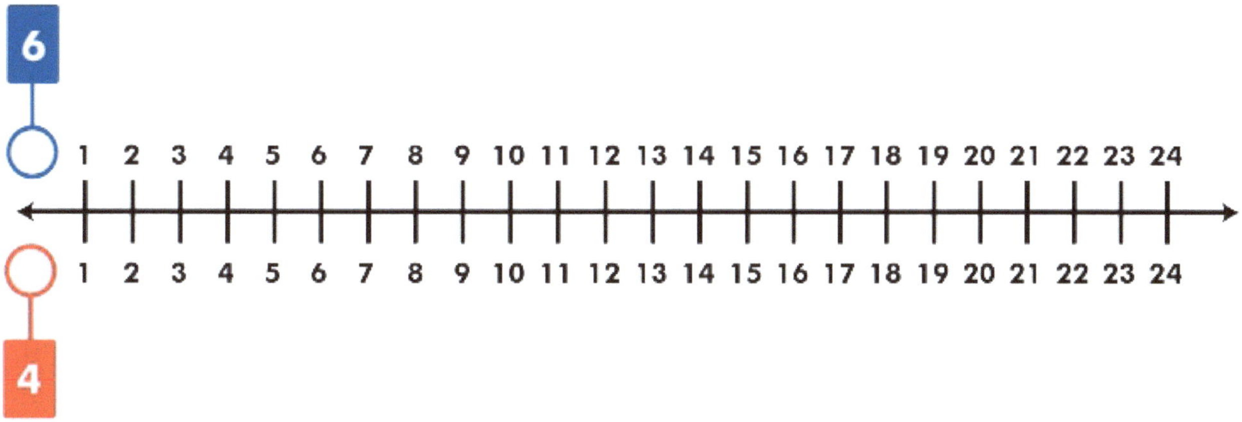

What are the common multiples of 4 and 6?

What is the least common multiple (LCM) of 4 and 6?

Using LCM to solve problems.

Draw a circle on the dates that Sanjeev goest to the gym and the the days that Amy goes to the gym. Use different colors for each student. Draw a green box on the date that they will meet at the gym.

Sanjeev goes to the gym every five days.

January

1	2	3	4	5	6	7
8	9	10	11	12	13	14
15	16	17	18	19	20	21
22	23	24	25	26	27	28
29	30	31				

Amy goes to the gym every four days.

When will Sanjeev and Amy meet at the gym?

When will they meet again?

Use the same method for identifying dates but start with their last meeting.

January

1	2	3	4	5	6	7
8	9	10	11	12	13	14
15	16	17	18	19	20	21
22	23	24	25	26	27	28
29	30	31				

February

			1	2	3	4
5	6	7	8	9	10	11
12	13	14	15	16	17	18
19	20	21	22	23	24	25
26	27	28				

Alarm and buzzer challenge.

The school's fire alarm is being tested and has been set to ring every 40 minutes throughout the day. The lesson buzzer sounds every hour. The school day starts at 8 AM and ends at 2 PM.

At what times during the school day will both the alarm and the lesson buzzer sound? Organize the images below to help you.

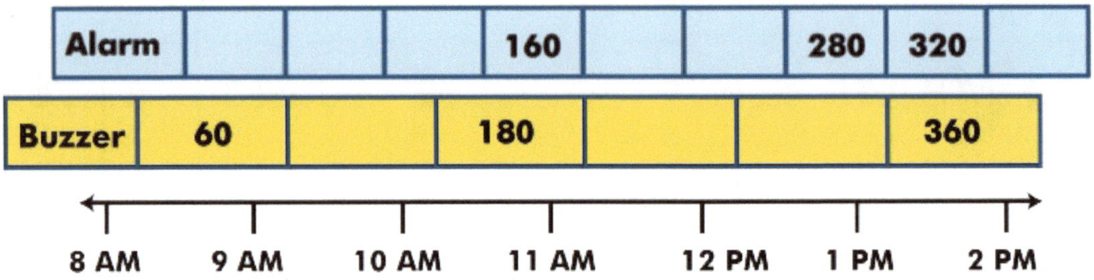

Name___________________________________

Multiples & Least Common Multiples (LCM) Quiz

1. 16 is a multiple of 4. True or False? ?

2. What is the LCM of 5 and 10? ?
   - A 5
   - B 10
   - C 50
   - D 15

3. What is the LCM of 3 and 11? ?

4. What is the LCM of 6, 7 and 14? ?

www.ingramcontent.com/pod-product-compliance
Lightning Source LLC
Chambersburg PA
CBHW052055190326
41519CB00002BA/235
9781630960766